I0821741

Mary Elizabeth Salzmann

Consulting Editor, Diane Craig, M.A./Reading Specialist

Sandcastle

An Imprint of Abdo Publishing
abdobooks.com

abdobooks.com

Published by Abdo Publishing, a division of ABDO, PO Box 398166, Minneapolis, Minnesota 55439.

Printed in the United States of America, North Mankato, Minnesota

052019
092019

Design: Christa Schneider, Mighty Media, Inc.
Production: Mighty Media, Inc.
Cover Photograph: Shutterstock Images
Interior Photographs: Shutterstock Images (all)

Library of Congress Control Number: 2018966947

Publisher's Cataloging-in-Publication Data
Names: Salzmann, Mary Elizabeth, author.
Title: Zoo Babies / by Mary Elizabeth Salzmann
Description: Minneapolis, Minnesota : Abdo Publishing, 2020 | Series: Animal babies
Identifiers: ISBN 9781532119620 (lib. bdg.) | ISBN 9781532174384 (ebook)
Subjects: LCSH: Zoo animals--Juvenile literature. | Animal babies--Juvenile literature. | Zoo animals--Infancy--Juvenile literature. | Zoo animals--Behavior--Juvenile literature.
Classification: DDC 599.039--dc23

SandCastle™ Level: Emerging

SandCastle™ books are created by a team of professional educators, reading specialists, and content developers around five essential components—phonemic awareness, phonics, vocabulary, text comprehension, and fluency—to assist young readers as they develop reading skills and strategies and increase their general knowledge. All books are written, reviewed, and leveled for guided reading and early reading intervention programs for use in shared, guided, and independent reading and writing activities to support a balanced approach to literacy instruction. The SandCastle™ series has four levels that correspond to early literacy development. The levels are provided to help teachers and parents select appropriate books for young readers.

EMERGING • BEGINNING • TRANSITIONAL • FLUENT

Contents

Zoo Babies

Can you find these baby zoo animals in this book?

baby alligator

baby elephant

baby hippo

baby koala

baby lion

baby
panda

baby tortoise

baby zebra

This is a baby alligator.

This is a baby panda.

This is a baby koala.

This is a baby hippo.

This is a baby zebra.

This is a baby tortoise.

This is a baby lion.

This is
a baby
elephant.

What Else Did You See?

branch

grass

sand

tree

Index

Teacher's Guide

ATOS: 0.8 GRL: A Word Count: 40

High-Frequency Words

a is This

Content Words

alligator, baby, elephant, hippo, koala, lion, panda, tortoise, zebra

Before Reading

- Tell students that the title of the book is *Zoo Babies*.
- Summarize the content of the book.
- Have students look through the book. Ask them what they see in the pictures.
- Choose a few new vocabulary words. Have students predict what letter each word starts with. Then have them find the words in the book.

After Reading

Ask students questions about the book's content, such as:

- What animals did you see in the book?
- Have you ever been to a zoo? What was it like?
- What other animals would you like to read about?